Amazon Fire TV Device:

Streaming Media Player with Voice Search

Candace Sinclair

Amazon Fire TV Device: Streaming Media Player with Voice Search

Printed in the United States of America

First Printing: May 2014

ISBN-13:978-1499682090

Disclaimer

The information provided in this book is designed to relay helpful facts on the subjects discussed. This book is not meant to be an exhaustive, step-by-step technical manual on how to use, troubleshoot, or otherwise make a buying decision for any products, services, or accessories, under any conditions.

The author and publisher are not liable for any damages or negative consequences from any action, application or preparation, to any person reading or following the information in this book.

Any references or information provided in this book are presented as the author's personal and professional opinions, suggestions, and observations and do not constitute endorsement of any websites or other sources. Readers should be aware that any websites listed in this book may change without notice.

Table of Contents

INTRODUCTION

I want to thank you and congratulate you for downloading the book, *Amazon Fire TV Device*. I hope you enjoy it.

This guide has been provided to help you understand the basics of the new Fire TV Device product created and sold by Amazon. By reading and reviewing this document, a user can learn more about the ways and requirements needed to use and interact with your favorite TV shows, movies, games, music, and so on from your Kindle Fire, TV, other devices, and your smartphone, via a streaming media device.

Amazon's effort for creating this wonderful Fire TV device is notable for various reasons, which are described here in detail. Apart from the features of the Amazon Fire TV device, this guide will provide you with an insight of the technical specifications so that you can make your best decisions when trying to choose a streaming video device that is right for you, your lifestyle, and the hardware that you have available to you right now.

CHAPTER 1: AMAZON'S LATEST GIFT TO THE WORLD OF STREAMING MEDIA

Marking a 350 percent growth in the world of video streaming service, Amazon finally introduced its own streaming device known as Amazon Fire TV. With this release, Amazon introduced tough competition to its competitors like Apple TV, ChromeCast, Roku, and many others in the battle for leading the marketplace in television streaming services.

This set-top device is a tiny box that can be connected to your HDTV so that you can enjoy Prime Instant Video, Netflix, Hulu-Plus, and low-cost movie rentals in an easy way. The Amazon Fire TV has the capability to combine all your favorite subscriptions along with streaming services into Amazon's amazingly massive selection of digital content.

With this product, you can now watch more than 200,000 TV serials and your favorite movies, rent videos for just 99 cents, and enjoy your favorite sports, music, news, and games, at no additional costs. What more can you ask for?

The Fire TV comes with a voice-activated search feature, and it can be doubled as a gaming console. Although cost-wise, it is the same as Apple TV, you will immediately appreciate the enhanced and brilliant attributes of your Fire TV device once you explore all its options.

According to Amazon's Vice President in charge of Kindle, Peter Larsen, they have hit it on the nail by creating this streaming device. This new Amazon product takes technology to the next level by overcoming the three weaknesses of existing streaming devices—search, a barred ecosystem of the existing streaming media devices, and slow performance.

To quote an example, Larson said that it is really frustrating for audiences when they can't watch their Prime Instant Video on their Apple TV. He sees the Fire TV media streaming device as the solution to the three listed problems. He claims the Fire TV device is three times faster than the other streaming devices offered for sale in the marketplace at this time.

In addition, the Fire TV interface is very simple to navigate. Straight out of the box, users don't need a training manual or a technology person sitting next to them in order to operate the device. Surprisingly, the moment you connect the cable to the back of your TV, you instantly gain access to all the additional Fire TV features that have intrigued Amazon's worldwide audience.

The Fire TV device includes features such as ASAP that predicts what the watchers might want to view next...sort of like an artificial intelligence sensor.

Currently, the apps that are available on Fire TV include Hulu, Netflix, WatchESPN, Crackle, MLB, NBA, YouTube, and Vimeo. Some of the attributes of Fire TV include the immense selection of things to watch, from your favorite TV serials, select HBO movies, and a range of more than 200,000 TV episodes and all-time hit movies.

Besides watching TV shows, HBO movies, and streaming video from your Prime Amazon membership, you have your choice of accessing millions of songs, games, and other favorites. The Fire TV is a fragment of a multibillion dollar effort by the company to move up the ladder from selling goods created by others to presenting the world with their own processes of consumption and creation.

In the world of books, Amazon has made its huge name by making e-readers and tablets along with the brilliant content it sells for them. Thousands, and maybe even millions of writers create the content that is exclusively posted on Amazon.

In the world of video, which is more competitive, Netflix has been the leader in providing streaming videos and creating original shows to be featured on a myriad of devices.

Streaming is the future of video, and keeping this in mind, Amazon evidently thought of giving tough competition to its co-fellows and the result came in the form of the Fire TV streaming video device that rivals many of the current leaders in technology. Amazon created this product after several years of hard work and deep speculation. Apart from providing the content from its own studios, the company has offered prime subscriptions to their retailers and customers.

Fire TV Voice Search Functionality

Talking about the promoted enhancements and areas of improvement, Fire TV comes with a voice search function and a prediction feature. Fire TV is expected to win hearts of millions and is definitely going to change the shape of online business and the way customers shop on the web.

With the introduction of this device, Fire TV viewers will be able to eventually use their remote control to buy any product of their choice from the web. Why? Amazon's goal is to combine its multimedia technology offerings with online retail businesses. The firm wants its customers to buy products and services from virtually anywhere, including the comfort of their living room. At the same time, biggies in the industry, like Apple, are struggling for space on the idiot-box, which is considered to be the home entertainment center for most families even today.

CHAPTER 2: HOW STREAMING AUDIO AND VIDEO WORKS

Media streaming is a method or a technique for transferring data in a way that it is processed as a continuous and steady stream. Streaming media can consist of video or audio content that is sent over the internet in a compressed form so that it can be played immediately rather than first saving it to any storage device like a hard drive. With the invention of this new technology, a user doesn't have to first install and wait for the download to happen in order to play a file.

As the media content is sent in a compressed format in a continuous stream of data, it can be played and viewed as soon as it arrives at its destination. Viewers can play, pause, rewind, and do everything that is possible with a downloaded file unless the content is still in the processing mode. This advancement is a technology that has taken the world of multimedia to its next level.

Streaming techniques have gained in importance with the growing need of anyone who has an internet connection, since most users' connections do not operate at a fast enough speed to download huge files while simultaneously watching or listening to the media at the same time. People do not want to wait for hours while a movie or audio file gets downloaded to their computers. We're living in a world of robotics where everything is expected to give us fast and easy access to whatever we want. With streaming, data can be displayed on the browser before the complete file is actually transmitted. It's like your dream come true!

In the early days of streaming media, watching movies online or listening to online audio pieces was not that much fun as it was like a stop-go, stop-go mechanism. On top of that, if you had a slow computer processor or a slow internet connection, you could spend hours staring at the buffering bar, which is as bad as watching grass grow or water boil. Even after struggling with incredibly slow download speeds, at best, all you got were jerky and time-paused videos. Those were not fun ways to spend your evening, whether alone or in the company of others.

From there, we have now reached an era where more than 56 million people listen to online radio each week. That's a tremendous change in the figures. This in its truest sense is called revolution.

Although the success of streaming media is just a newborn baby from where technology will take us in the short years ahead, the idea cropped up in the minds of experts ages ago. We all know that information travels to us in the form of a wave. The data carried in this information wave is decoded by our ears and brain, which is then understood by us.

In a similar fashion, information travels to an electronic medium in the form of a cable or radio waves, and then the electronic device decodes it from where it is displayed on our screens. This forms the basis of streaming media.

How Does Streaming Technology Work?

This is the question that might be bothering most of you, especially after learning its pluses. For the streaming technique to function effectively, the client side [the device where you will watch the streaming video or audio] that is to receive the incoming data should be capable of collecting it and sending it as a continuous and steady stream. This continuous data is sent to the application that processes it and converts it into the required audio or video. This means that your TV, iPad, Amazon Fire Kindle, or other media device is expected to receive data at a very high speed so that it can save excess data in the buffer. If the streaming client fails to receive the data quickly, the entire process will slow down and the production of data will not be as smooth as it is expected to be.

Media files are normally streamed from pre-recorded or pre-created files, but sometimes, these are also distributed in the form of a live broadcast feed from several types of live sources.

So, how does it work in broadcasting? In a live broadcast, the media signals are converted into compressed digital signals, which are transmitted from an internet server to multiple users, at the same time, in the form of a single file. The media data is transmitted by a server system that is received and exposed in real time by the client system known as a media player. This media player can either be a part of a browser or a separate entity, a plug-in, or a dedicated device like an iPod. Normally, video files come with their compatible-embedded players, such as the YouTube videos you've watched that run in flash players.

Pros of Streaming Media

Several major advantages of streaming media, which are gaining popularity include the following:

- The technique makes it possible for its users to make use of the interactive applications like playlists, video search, and so on.

- It provides an effective and efficient usage of bandwidth, because only the section of the file that is being transferred is the part that is being watched by the users.

- It lets the content be delivered to the user's screen at the exact time they are watching it, in a smooth and quick fashion.

- It saves lots of space in memory as the file that is being viewed is not saved anywhere on the storage device. This also provides more control to the content creator as the file that is being seen by the user is not stored anywhere on a local drive.

Apart from these advantages, you might want to learn about the other pros of using streaming media after the first few times that you have enjoyed using it. Streaming technologies have greatly improved since the time they were first introduced, except for the fact that the quality of audio or video still depends on the connection speed of the internet.

Chapter 3: Enjoying the Brilliant Features of Amazon Fire TV

Amazon is in the hit list of business. Although the company never reveals its sales figures, it has gifted the world with some of its best sellers. Amazon's tablets and e-readers have rocked the market with some of their amazing features. And now, it has hit the stores with its third best seller—the Amazon Fire TV, the new streaming media device, and it has been created by no accident.

Amazon has quoted the selling price of its new creation as $99, which gives competition to other set-up boxes like Roku, Apple TV, and Google ChromeCast. Fire TV is expected to live up to the expectations of its users, who want to cuddle up in their couches right in front of the big TV screen. The brand has driven huge resources behind its streaming capability, streaming – which is the primary method of operation on the Fire TV. The company claims that it is going to make money not when you buy its new device, but when its customers actually use it.

After learning the facts and workings of streaming media, we can now have a look at the salient features of Amazon Fire TV, which makes it "Neighbors' Envy; Owners' Pride." The box comes with various internet video features that you would expect from a modern set-top box, apart from some other new ones, for instance, the predictive feature. Amazon seems to have nailed them all, and to help you decide if the product is worth its cost, below is a quick summary that lists all the major functionalities of the Fire TV.

Features of Amazon Fire TV

It's a tiny box with great specifications. The product comes with a fast quad core processor with 2 GB of memory and a dedicated CPU. Apart from this, it has a 1080p HD video and Dolby digital sound system. It comes with 2 GB RAM, which is four times the amount present in the biggies like Roku, Apple TV, and ChromeCast. You can now enjoy videos and audio with a processor that is three times more powerful than its competitors, and offers customers instant search results via voice commands to the responsiveness of the interface, from the smooth gaming speed to the supersonic speed of streaming.

Everything and every feature of this tiny box is superfast. The credit goes to the super fluid quad core processor. Amazon's brilliant product employs Advanced Streaming and Prediction Software, which has the capability to dynamically learn which shows and movies you are fond of, and it gathers suggestions in the background for you to choose from in an easy-to-access "watch list."

It's based on Amazon's existing technology that is available on their site. For example, when you buy or download a Kindle book, Amazon shows you other books that you might be interested in, based on your selection and purchase. The same is true for the way that the Amazon Fire TV device works. It has not left a single chance to spoil its customers to get a break in those few seconds when you're thinking about what to watch or listen to next.

To be precise, "ASAP" seems to be a feature that can predict what you would want to watch next. It provides True Instant watching, in the real sense. This is the feature that has marked a huge improvement for Amazon's device over its competitors. So, the more you use the TV, the more accurate the feature of ASAP becomes as it dynamically adapts to your watching habits.

The ASAP feature not only predicts what you would want to watch, but it also gets the video piece ready for streaming by intelligently starting the buffering process.

What The User Needs to Know About This Feature

The ASAP feature is not 100 percent accurate all the time, but it learns from your watching habits over a period of time. This can be seen as a very convenient streaming option as it speeds up the entire process while you are watching your favorite TV show. But people who are quite particular about their privacy might not want it to make note of their watching habits, because, in the background, this technology is electronically documenting your entertainment choices.

So Many Choices, So Little Time

At your fingertips is an enormous selection of videos and audios, which is another salient feature of the Fire TV device that offers its users over 200,000 TV series choices and movies, millions of online songs, and thousands of games. It brings your small screen to the bigger screen by using this amazing feature.

You can mirror your tablet on HDTV and share movies, videos, and TV shows from your tablet with everyone else sitting in the living room, on the big screen. What you have on your Kindle Fire HDX is what you see on the HDTV. You can control the mirroring through one simple command to let it know when it is supposed to start the mirroring process and when it should stop. That's amazing!

Fire TV is Compatible with Amazon Prime

There has never been a better way of enjoying Prime Instant Video, because with Amazon, it is always a prime time. Amazon Fire is the best technique to experience the brilliant Prime Instant Video. There are various benefits that are offered to prime users; prime members are free to use unlimited and commercial-free streaming of millions of popular TV shows and all-time favorite movies. Prime customers are free to enjoy all of these with no additional costs.

Just say what you want to watch and Amazon Fire will make your dream come true. The product comes with a voice search feature that actually works. Just speak the name of the TV serial or a movie that you wish to watch, and you can start watching it within a few seconds. This feature lets you search less and watch more. Just speak into the remote and tell your TV what you want to watch. The remote of the product comes with an in-built microphone, which you can use to browse through various movies and TV shows belonging to various genres. It brings the world to your fingertips.

After surveying their customers, Amazon learned that browsing through the giant music and video libraries was one of the most annoying concerns for anyone owning a streaming media device. So, Amazon pinpointed this frustration and accelerated the product's success by offering voice search in an instant. This has made searching tasks pretty easy for couch potatoes.

What the user needs to know: The statement made by the company to fulfil its users' dream is partially true. Although there is a dedicated button on the remote control device to control the browsing option, the feature is not compatible with all the apps. It has been tested to work with any subscriber who has a paid Hulu or Netflix account, but the company claims it will be incorporated into other apps in the near future.

Amazon Fire TV Device is User Friendly and Easy to Install

Amazon Fire can be preregistered to your existing Amazon account so that you do not have to spend hours in registering it or configuring it. When you open the package, you will quickly learn that it has an easy navigation system and powerful recommendations that are some of the salient attributes that makes the product very user friendly.

The simple-to-follow interface makes it easy for you to search through various available options and provides you with personalized recommendations, keeping in mind your choices and habits. You can add more choices to your Watch List which is actually the Wish List. Talking about the remote, it is quite simple to access and is equipped with all the controls you need to search, browse, watch, and play games. It uses the technology of Bluetooth to connect to the Fire TV, so just relax and enjoy!

Enjoy Instant Streaming Media with the ASAP Feature

With the latest feature "ASAP" that has the intelligence to predict what movies and TV shows you might want to watch, you can enjoy the instant streaming of data as the TV buffers the data and keeps it ready in advance of when you need it, and it's available the second you hit the play button.

Fire TV is Great for Gamers

If you are a gamer and love spending time exploring new gaming areas, this product can provide you with millions of games. For gaming, Amazon TV customers will now be able to play on the small remote or a tablet, or with a beautiful Fire game controller. Amazon now lets you pay less and play more with the most affordable and easiest way to play your favorite game on HDTV.

Whether you are an individual or have a family that likes The Game of Thrones or you enjoy car racing games, Fire TV is an amazing way to play and bring these favorites to you within seconds. Enjoy these games with the remote or a dedicated game controller, and just know that this device is especially designed for your needs to go deep into the action mode.

Amazon has tried to go beyond the Angry Bird and other racing games. The set-top box offers its users side-scrollers, and many other interesting games. These games are also equipped with the multiplayer option mode. For playing these games, you can either choose to use the set-top box remote control device or a separate Bluetooth device, especially for playing purposes.

What the user needs to know: The TV comes with all these games that have Fire TV specific versions like Gameloft's Asphalt 8, Telltale Games' The Walking Dead and so on. Amazon has its own gaming studio, which is built from scratch especially for the Fire TV. For the multiplayer games, you can hook up with your friends using mobile devices on Wi-Fi. In such cases, there will be one primary user who will determine who can make use of a Bluetooth connection device for a better and direct connection.

Fire TV is Perfect for the Family

Amazon's FreeTime enables you to limit the screen time and create some personalized and customized profiles that help you set the needs and demands you want to allow for your kids. This features lets its customers integrate the device with X-ray. In other words, it allows its users to access information on a Fire HDX tablet. We all know that there is a lot of adult content out there on multimedia that parents might not want to expose in front of their kids and teenagers. So, to ensure this, the FreeTime app was redesigned to give a kid-friendly and dedicated zone to all the families who own this Amazon Fire TV device.

What the user needs to know: If parents have already configured their Kindle Fire tablet to make it kid friendly, the settings can be automatically synced to the Fire TV device. There are password protected profiles, and parents can have the timer set to manage the screen time for their kids. It will help them in communicating when it is time for them to watch streaming media, and when they have to go for their homework. But this feature can be enabled with the required subscription. The FreeTime Unlimited has all these features along with unlimited access to TV serials and movies.

Amazon Fire TV is Made for Music Lovers

If you are not into the gaming world, but you are a big-time music lover, Fire TV has something in store for you too. You will find millions of tracks available to stream to the Amazon TV through TuneIn, Pandora, and iHeartRadio.

In addition, you can choose to access your complete Amazon music library, so that you can enjoy the songs and albums you previously purchased from the Amazon store. You can play games while listening to music or read news with light music in the background. Fire TV is there to tell you what is playing as soon the new song starts.

A First-Class Second Screen Experience

You might be wondering what this Second Screen Experience actually is as this has become quite a famous phrase among media experts who love to toss around the term. What they are trying to refer to is the act of looking up on a smartphone for data that is quasi-related to what you have on the TV. Fire TV's X-Ray feature has automated this process for you.

If you own a Kindle Fire HDX and once the sync-up activity has been performed, the X-Ray will pull all the relevant information from IMDB (Internet Movie Database) – information like crew details, star cast, reviews, story line and so on. Have you ever wondered who the star actor was in the movie that you were currently watching? Or what the name of the song is that you are listening to? Well, you do not have to boggle your brain any more to find answers to all such questions, because X-Ray is now there to help you.

What the user needs to know: This X-Ray feature right now works only on the Amazon Fire TV that comes with Kindle TV HDX, although the support is going to be extended to iPhone and iPads very soon.

Streaming Alternatives

When you are confused between the various shopping options, let the Amazon Fire TV be your best buddy to guide you to pick the best choice. When you reach any show, the Fire TV features all the available streaming choices on the screen so that it becomes easy for the shopaholics to pick the best product for the cheapest price.

What the user needs to know: This feature of the Fire TV is not compatible with all the services. The only services it currently works with are Prime Instant Video and Hulu Plus, apart from its own Amazon Instant Video. But the company promises to extend this feature to other providers as well.

Casting

The fact is that no streaming media device can be tagged as a complete package without casting. Keeping this in mind, Fire TV users can cast the streaming media from their mobile phones to the television while using these phones or tablets as the remote controlling devices.

What the user needs to know: As of now, the feature is only compatible with the Kindle Fire HDX and soon will be extended to the iPhone and iPad. Kindle tablets can also display their content on the HDTV.

With all these and many other features to come, Amazon knows that its customers on Kindle spend more time and money on Amazon than other non-users, so the streaming device is a gift for them so that they can utilize their streaming content.

Let us have a look at a few important variables that can make or break the way for a new launch in the electronic world.

- **Cost**. Amazon Fire TV is going to be tough competition to its co-fellows as it makes much more sense for a brand like Amazon to have its own hardware so that its users can utilize the amazing streaming content in a better and compatible manner. Its price is much less than other biggies like ChromeCast.

- **Power**. Does any single electronic box have all the powers in it? Well, it sounds like an impossible thing to have, but the new Amazon Fire TV has it all. Apart from the fact that the Fire TV can stream video over the web and thereafter onto your TV box, Amazon has made all the hard work easy to prove it to the world that the Amazon Fire TV is powerful.

- The new set-top box is used for two purposes: searching and browsing. With the use of a better quality processor, searching becomes quick as compared to other competitors like Roku and Apple

TV. Some of you might be wondering why browsing is such a challenging factor when comparing one device to another, but it matters a lot when you're searching through the tremendously large libraries. By normal standards, it's not only time-consuming, but it's also frustrating. But with the new voice search option, Amazon's Fire TV device solves all those irritating problems.

Chapter 4: Comparison with Co-fellows in Competition

In order to understand the beauty of this new product by Amazon, it would be interesting to compare and see various features with its competitors like Apple TV, Roku, and Google ChromeCast. Google, Amazon, Apple, and Roku are wrestling constantly to win the trustworthiness of cord-cutters. Lucky are the viewers who are benefitting from the advantage of this fight as they are getting streaming media in the compressed and continuous manner right on their HDTV box via the means of internet.

Let's take a look at the facts.

- **Price**. Comparing the price of various models, Amazon Fire TV is at par with the Apple TV, but leading in the league is Google ChromeCast, which is selling the product for a cheaper price. Roku takes the second place after Google. But each of these devices cost not more than $100, which doesn't include extra features like gaming controllers and so on. We know that Roku offers three different versions from $50 to $100 for Roku3. However, the point that should be noted here is that the cost of none of these models includes the cost for online video on demand or services like Hulu Plus, Netflix, or Amazon Prime. There are separate subscription charges for these services. But Amazon offers its services for a good amount to its loyal customers so that they can make the best use of the content on the Amazon hardware.

Although there are extra charges for additional features, the cost that you are saving for a Dish Network is much more than what you are spending to buy the subscriptions. Even if you take the case of games, subscriptions to some of your favorite games will still cost you more money for several years to come. So, overall it is a win-win condition for the users.

- **Shape and size**. All the streaming media devices are essentially available in two formats: the wallet-sized set-top box like Amazon Fire, Roku, and Apple TV, and the thumb drive like Google ChromeCast and Roku Streaming Stick. But the common feature about these models is that all of them bank on Wi-Fi and an internet connection for functioning. If you compare your old conventional box with these modern devices, you can see that it looks like a piece of antique technology. Except for the Amazon Fire TV Game Controller, there are no compatible extras for other models.

- **Content**. Content is one of the most crucial attributes to base our comparison among the various streaming media devices. The biggest difference between these modern devices lies in the way and the type of content these products provide us to access. Biggies like Disney, ESPN, and a few more are available on almost all of these boxes but others differ.

For instance, Amazon Fire TV provides access to HBO but not VUDU. So far, ROKU is known to have the highest number of content providers, but when you look at the list, you might feel that you are not aware of most of them. ChromeCast is known to have the least number of providers, but the number is expected to grow with time as Google has now come up with the feature that it allows other technologies to get merged with its apps.

Amazon takes the prize by distinguishing itself with a huge variety of games, as this is the feature in which other devices lack. Roku, although, offers various simple games that can be played using its remote. All these brands let you buy the content via their associate stores and are designed to drive it on rent. Roku is slightly different as most of its channels are free. But some of these channels require an amount as subscription fees before gaining access to the desired media.

- **Performance.** Taking performance into consideration as the factor to compare various media devices, Apple, Roku, and Amazon have a pretty user-friendly interface to operate and uses various navigation options. They all come with a slim, easy-to-handle remote control too. But Google belongs in a different league altogether, if we talk about performance. You can control the navigation options with your mobile device, and that is how it integrates the control signals with its specific mobile apps like Pandora. However, you need to have a smartphone or a tablet in order to control the options, and to act as a user interface, since there is no other central interface provided. What a different animal!

If you are a hardware fan, this brilliant and tiny Amazon Fire TV box is sure to entice you with its fantastic attributes. You can think of it in this way: Fire is not an animal to provide minimal hardware; it isn't a product with basic casting features for a cheap price tag, and it goes way beyond all these elements. Fire TV is much more powerful with mind-blowing specs and more features than other hardware at almost three times the price.

Amazon Fire TV comes with a feature of voice search, which is the most unique feature among all the other set-top boxes. Like the conventional search option of Roku, the voice search of Amazon lets you browse through various video services in order to find the best price for a required TV show or a movie. The only point to keep in mind here is that it works only with Amazon's own videos and Hulu Plus.

If all you want is to limit or control the media search options especially for the kids in the house, Amazon lets you do that also with its special feature called the FreeTime parental control. This features lets you have a special profile for each child in the house, and you can determine what he or she is supposed to watch, with limited control for the various media options.

Roku also provides you with a control on these options; you can use a PIN to limit the options and to control the number of newly added channels. Even Apple TV lets you control the various features using a special PIN.

Apple TV claims that its device is three times faster in performance than other competitors in the battle. Although the required tests have not been performed to check the authenticity of this statement, it is suspected that the user's video watching know-how depends a lot on the speed and consistency of the internet broadband connection and the Wi-Fi system that is present inside the device.

Comparison Conclusion

It can be concluded that the Amazon Fire TV is much more tempting than all the other competitors in the same league. However, the device doesn't come with the required options that allow it to overcome the two biggest hurdles that would bid goodbye to Dish connections and other streaming media devices. What am I referring to? The Fire TV lacks the broad selection of news and sports. Apart from this, if you were to make a decision of picking up one of the mentioned devices, the selection depends a lot in part of where your product loyalties belong.

If you are a loyal customer of Apple, you will surely find the Apple TV to be much more easy and simple because of its unbroken integration with iPhone and other devices.

If you are a fan of Google, Chromecast is surely going to tempt you as it has a tougher appeal for people who love everything about Android and are not bothered about millions of video options.

COURTESY: GOOGLE

For all the moms, Roku is a good option as it is a simple box with all the needed features. Kids, on the other hand, will definitely be tempted to pick a Fire TV as they would be attracted with the super cheap library of the games.

Chapter 5: Amazon Takes the Streaming Media Lead Over Others

In today's streaming world, where lots of high-tech companies want to sell their streaming media devices to users to watch movies, shows, and video, it is this author's opinion that Amazon takes the lead with its new creation of the Amazon Fire TV streaming media device. Amazon made it very clear why it created this new set-up box. They saw that most users were not happy with the set-up box that interacted with the HDTV, and they were not really sure of the plentiful data that these devices could deliver. Amazon didn't do any great discovery in creating this hardware; there were no new innovations around novel product categories, like other competitors. Also, they do make a profit by selling their own hardware.

The only thing the brand has done is that it takes all the data, data that it gathers being the world's top retailer, looks at and observes what is obtainable and by looking at what consumers want, it creates a piece of hardware, a small device that it can sell for a cheap price just to get its users into its network.

To quote examples here, Netflix created House of Cards to satisfy the growing needs of a group of viewers. Similarly, Amazon created the Fire TV because it could see the growing need of a fantastic streaming media device in the market. Among its competitors we have Google's Chromecast, two Roku models, and the Apple TV that are in the top categories.

Amazon plans to make money by using Fire TV as a tool to sell everything else from its repository; this is how the brand has always worked towards stepping up the ladder of success. It gives its users various options, and then depending on their needs, uses these options to provide more features. According to the creators, once its potential buyers have the FIRE in their living rooms, they will see to it that the users buy their content, movie libraries, and TV shows. A few customers might want to sign up for Prime. Hence, Amazon tries to make sure that its customers make the utmost use of stuff that they buy from the store, not just buy it to keep it in their homes.

This strategy is followed by all three hardware products of Amazon, especially their tablets. Amazon knows what people want so they market their products well in the immature market. They promote them effectively on the web, and especially on their powerful website. It also gives them another benefit; when people visit their website, some of them make a choice to order other products. So, wouldn't you call this "a perfect marketing strategy?"

With the introduction of the Fire TV device, the company has put the Prime experience in front of their audience in a way that no one else has done so far. The tactics Amazon uses to understand the market's needs are that it studies their audience's shopping habits and their needs. Then, accordingly, it gives them what they want in a way and speed that they demand.

Amazon gives their customers the products and all the other features that are required to use this new product. And I'm sure you would agree that Amazon has built its empire on its integrity. What does that mean? Consumers trust Amazon to deliver quality products, and they have a refund and return policy that is unequaled in the world. In addition, if you purchase a product from Amazon, they even allow you to resell it on their website, AND they pay you for the shipping fees. Now, honestly, what other companies in the world do that?

We've heard of users who complain about the slow processing speeds of Roku, yet Amazon offers to give them the fast speed Fire TV. Consumers complained of how difficult it was to search for their favorite shows or actors, but then Amazon created a product with voice-search options in their Fire TV device. Amazon doesn't have to wait to understand what people want. It just goes with their reviews on other devices and other brands. Intelligent move!

Similarly, before coming up with the new set-up box, Amazon was watching the market and its customers' reviews for quite a while. Then the company started to grow the seed of development by working on the problem: "Do we want to actually make use of the boxes that are lying dormant in someone's living room?"

This was the question on the usage of the other streaming media devices as people were not too sure about using them. It was probably at that time when Amazon prepared a checklist containing the demands and needs of the people. But as we've all come to realize, it wasn't just a checklist; it was a success list for Amazon.

CHAPTER 6: TIME FOR SOME TECH DETAILS

Talking about the technical details of the Amazon Fire TV, let us have a look at the attributes like size, weight, storage, and so on.

The Amazon Fire TV comes in the defined size of 4.5" x 4.5" x 0.7" (115 mm x 115 mm x 17.5 mm) and it weighs less than 9.9 oz. (281 grams). The SOC platform that is being used by this streaming media device is the Qualcomm Snapdragon 8064. Fire TV makes use of the superfast processor named Qualcomm Krait 300, which is quad-core to 1.7 Ghz. It is available with a storage capacity of 8 GB internal and memory specifications are 2 GB LPDDR2 @ 533 MHZ.

Another great feature of the product is its Wi-Fi connectivity. The product is equipped with a dual band dual antenna Wi-Fi, which is typically provided for the faster streaming mode and less number of dropped connections than all other standard Wi-Fi connections. It supports both the public and private Wi-Fi networks that in turn use the 802.11a/b/g/n standard with support for WEP, WPA and WPA2 security using password authentication.

The Bluetooth version that is available with the product is Bluetooth 4.0 with support for the following profiles: HID, HFP 1.6, and SPP.

Users will be amazed at the fact that the Fire TV has free cloud storage for all Amazon content. Again, this proves the validity of a statement made by Amazon that it makes its profit by buying everything that is required to make people use their products.

Ports. It comes with a 5.5 mm DC Jack Type A HDMI 1.4b output, w/HDCP Optical Audio (TOSLINK) 10/100 Ethernet USB 2.0 Type A.

Audio support is another great feature as the device comes with support for Dolby Digital Plus, 5.1 surround sound, 2ch Stereo and HDMI audio pass-through up to 7.1. The formats that are supported are:

Video: H.263, H.264, MPEG4-SP, VC1,

Audio: AAC, AC-3, E-AC-3, HE-A, PCM, MP3

Photo: JPG, PNG

For the output, the device demands a system requirement of high definition television and an HDMI cable. It is compatible with high definition TVs with HDMI capable of 1080p or 720p at 60/50Hz, including popular models from these manufacturers: Hitachi, JVC, LG, Mitsubishi, NEC, Panasonic, Philips, Pioneer, Samsung, Sharp, Sony, Toshiba, Vizio, and Westinghouse.

Not to forget is the fact that the device comes with a one-year limited warranty and service is included (terms and conditions apply). Although certain services may not be available outside the U.S., they provide full support for available content on applications which provide closed captioning functionality.

Technical Specifications of the Fire TV Remote

The Amazon Fire TV comes with an Amazon Fire TV remote that comes in the size of 1.5" x 5.5" x .6"(38.3 mm x 139.9 mm x 16.1 mm). This beautiful slim remote weighs less than

68 grams or 0.15 lbs. with batteries (45.5 grams or 0.10 lbs. without batteries). The device uses two batteries of AAA size that are included in the package. The remote control device works on Bluetooth technology that is Bluetooth 2.1 + EDR with support for the following profiles: HID, HFP 1.6, and SPP.

In order to make sure of the Voice Search Option, the device is equipped with dual digital microphones with noise suppression technology integrated. Altogether, there are all the buttons that are needed by a customer to peacefully cuddle in the couch and let the remote device do all the necessary functions. Hence, there are buttons for Voice, 5-way Directional, Back, Home, Menu, Rewind, Play/Pause, and Fast Forward.

What Comes in the Box?

The other items that you get as part of the Amazon Fire TV package include the following:

- Amazon Fire TV remote control device
- Power cord
- 2 – AAA batteries, and
- Quick Start Guide

CONCLUSION

With the announcement of its new product, Amazon has taken the world of streaming media to the next level. The Amazon Fire TV is expected to face the other giants wrestling in the same battle – to name these competitors, we have Google ChromeCast, Roku, and Apple TV.

This guide has been created to highlight the various features that this Amazon Fire TV streaming media device has for its users. These features were introduced keeping in mind their belief that the company earns profit not when its users buy the product, but when they actually use it. To ensure the validity of this statement, Amazon has linked several features to its existing content so that their users can benefit the most from the Amazon experience.

Although there is no end to the choices of streaming media devices available in the market today, it appears that Amazon has tried to throw everything else in the wash basin with the introduction of its new, tiny, set-top box. We believe that Amazon probably hopes that its product will entice its users for a long time with all the live streaming options both now and in the future.

ABOUT THE AUTHOR

Candace Sinclair delights in finding, testing, and writing about new technologies and those passions in her life that she likes to share with others.

She lives on Whidbey Island, Washington, has three grown children, four grandchildren, and she delights in exploring the Pacific Northwest through her photography and teaching others about becoming profitable published authors.

To learn more about what she's up to, and to download free reports and photos, you can visit her website and blog at: http://candacesinclair.com

DID YOU ENJOY THIS BOOK?

Thank you for purchasing this Amazon Fire TV Device guide. If you'd like to leave a review on Amazon that tells others how they might benefit from reading this book, you'd make me quite happy. Thanks!

www.ingramcontent.com/pod-product-compliance
Lightning Source LLC
Chambersburg PA
CBHW071812170526
45167CB00003B/1279